Oliver Joseph Lodge

AF812157

Hurter Memorial Lecture

The Structure of the Atom

Elibron Classics
www.elibron.com

Elibron Classics series.

© 2007 Adamant Media Corporation.

ISBN 0-543-70089-5 (paperback)
ISBN 0-543-70088-7 (hardcover)

This Elibron Classics Replica Edition is an unabridged facsimile of the edition published in 1908 by Vacher and sons, London.

Elibron and Elibron Classics are trademarks of Adamant Media Corporation. All rights reserved.

This book is an accurate reproduction of the original. Any marks, names, colophons, imprints, logos or other symbols or identifiers that appear on or in this book, except for those of Adamant Media Corporation and BookSurge, LLC, are used only for historical reference and accuracy and are not meant to designate origin or imply any sponsorship by or license from any third party.

REPRINTED FROM THE JOURNAL

OF THE

Society of Chemical Industry.

LIVERPOOL SECTION.

JULY 31, 1908.

HURTER MEMORIAL LECTURE.

THE STRUCTURE OF THE ATOM.

BY

SIR OLIVER LODGE, D.Sc., F.R.S.

LONDON:

VACHER AND SONS, WESTMINSTER HOUSE, S.W.

1908.

(17068)

[*Reprinted from the Journal of the Society of Chemical Industry, July* 31, 1908. *No.* 14, *Vol. XXVII.*]

Liverpool Section.

Meeting held at the University, on Thursday, March 12, 1908.

DR. J. T. CONROY IN THE CHAIR.

HURTER MEMORIAL LECTURE.

THE STRUCTURE OF THE ATOM.

BY SIR OLIVER LODGE, D.SC., F.R.S.

The atom of matter, sometimes called the ultimate atom, has been the object of frequent speculation, and likewise the subject of much scientific scepticism. In my youth, chemists usually guarded themselves, when they used the term atom, from being supposed necessarily to imply that such things really existed. And when they depicted a possible arrangement of atoms in a molecule, by means of a graphic formula, it was always in an apologetic spirit, as if such representation conveniently expressed some part of the facts, without necessarily representing anything really existent in nature. This scepticism was in some sort wise, no doubt, but in my judgment was carried too far. If a graphic formula conveniently represented some of the facts, it must be because it genuinely corresponded to something real and true. That it was a complete or adequate representation no one could imagine; but that it was worthy of being favourably regarded for whatever was true in it, without being subject to constant detrimental criticism, I fully believe. How *could* the formulæ correspond to reality when they were only diagrams drawn upon flat paper? They certainly could not; but these drawings might have been regarded as plane *projections* of a grouping in space, and an effort might have been made to conceive the real groupings which could be thus imperfectly represented. Had there been faith in the reality and genuine character of some such grouping, the attempt to realise what it might be really like would have been made far sooner than it was, for the idea of such an attempt is an obvious one;—the merit of those who have now at last properly worked at it (Van t'Hoff and others) consists in the way they have elaborated the details and applied them in a consistent manner to explain the properties of a great variety of compounds and their allotropic forms.

So also the bare idea of "survival of the fittest" was an obvious one—the narrative of every shipwreck and most fires and colliery disasters contain examples of it; the idea and its tendency could be, and were, conceived clearly by a naturalist in the space of some two hours. The major merit is held to belong to the careful elaboration of details, and the patient scrutiny of theory in the light of a vast number of facts, over a period of thirty years. But this work would not have been undertaken if there had been fundamental scepticism as to the existence of species. The existence and problem of species was the stimulating factor.

The old idea of species has been modified, however—there is nothing absolute about it now, as there was when each was thought to be a special creation. And similarly the idea of an atom has changed—there is nothing ultimate about it now, as there was when it was thought that the atom of each element was a special creation. The notion of evolution has invaded both departments of knowledge; and atoms are seen, in some cases at least, to merge into one another, and to be resolvable into other and still simpler units. But, in spite of all that, the atom always was and shall remain a useful conception; and the arrangement of atoms in a molecule, as now depicted by chemists, in all probability corresponds with something quite genuinely real. I see that Prof. Larmor suggested the other day, in Manchester, that such arrangement may directly determine crystalline form; and that a study of crystals may be one of the clues to real molecular structure. This notion should be pressed and examined for all it is worth. The fear lest any hypothesis may become too fixed and binding is wholesome, but it goes too far when it exerts a deleterious influence on thought and discovery. If we could dissect out and examine a molecule, it is highly unlikely that we should find anything static—a frozen rigid assemblage of atoms grouped in the way we depict; but it seems to me extremely likely that the actuality, when we find it, will kinetically resemble and correspond to so many features in our present static representation that we shall have to admit such representation as a genuine step in the direction of truth. And a step is always a thing to be *taken*, not to be criticised and boggled over. Each step upward, if properly taken and utilised, makes the next step on more accessible and easier. And if it happens to be a false step, too weak to bear any weight, we shall ascertain the fact most quickly by frankly treading or even jumping upon it.

At present we are dealing, however, not so much with the chemical structure of the molecule as with the physical structure of the atom.

This lecture is given in memory of our friend the late Dr. Hurter of Widnes, who among all the technical or industrial chemists of my acquaintance stood out as one who earnestly strove to realise with his mind's eye the inner working of things beyond even the aided organs of sense; and who, whether in the study of scientific photography or of electrolysis or other parts of physical chemistry, endeavoured to picture to himself as clearly as possible *what* it was that was really happening in the atomic and molecular domain. He gave two memorable papers to the Physical Society of the University College on these subjects, and his patient and persevering search for accurate knowledge must have impressed everyone who heard him.

Attempts to visualise the actual processes occurring in nature are wholly legitimate, and if persisted in long enough are nearly always fruitful. It is the glory of science to take what is superficially complex and empirical and confused, and to convert it, by long-continued critical and imaginative inspection, into something orderly and systematic and intelligible; probing into the hidden nature of the working, and disentangling from the obvious or phenomenal, that which, though unperceived by the senses, is really far more accessible to mental grasp. Every such generalisation is a step in the direction of unification, and leads us on toward that ultimate unity, and in a sense simplicity, which is the object of all philosophy and of a great part of science. For the multifarious appearances around us are probably all going to be resolved into the movements and groupings of an immense number of precisely similar elements. Our business is to ascertain what those elements are, how in each case they are grouped, by what means they are connected, and how they act upon each other. And then we may further inquire how far they are influenced by some other aspects of the universe, beyond the material range, outside the familiar categories of matter and ether and electro-magnetism and mechanics.

Chemical Affinity.

One of the salient discoveries of the last century, on which sufficient stress is seldom laid, is the discovery that chemical affinity and electrical attraction are one and the same—that the electric force between charged bodies is really chemical action at a distance, and that the chemical combination of atoms and molecules is due to their electrical attraction. Even what is called residual affinity, or latent valency, or molecular combination, can be thus explained also; for the "bond" of the chemist is thus seen not to be an indivisible unit, but to be made up of a bundle of lines of force, some of which are available for taking hold in a number of minor directions. And thus the phenomenon of "mass-action" becomes completely intelligible. And in this same way the at-first-sight surprising dissociation of salts, by the mere fact of dilute solution, is explained likewise. The fact of physical cohesion, the linking together of saturated chemical compounds into a mass with mechanical tenacity, the whole subject of "strength of materials" in fact, seems explicable also on this doctrine of residual affinity or subordinate electrical attraction between molecules. For tenacity, though great enough in the aggregate, is a small force per unit, and is incomparably smaller than the forces specifically known as chemical. Just as gravity, though eclipsing them all when planetary masses are concerned, is trivial when compared with cohesion for bodies of ordinary size, and is practically infinitesimal among atoms and molecules.

For a first instance of this, take the gravitational attraction of earth and moon: the force needed to hold the moon in, while whirling round the earth once a month, would tear asunder a steel rod 400 miles thick. For a second instance take the tenacity of 30 ton steel: holding itself together with a force of 30 tons weight per square inch, or, expressed in c.g.s. measure,

$\frac{30 \times 10^9}{6 \cdot 5}$ dynes per sq. cm. Reduce this to the force across a molecular section, and it becomes $\frac{5 \times 10^9}{10^{16}} = 5 \times 10^{-7}$ dynes per pair of molecules. But if these molecules were electrically charged, and operated likewise at molecular distances, the force between them would be $\left(\frac{3 \cdot 4 \times 10^{-10}}{10^{-8}}\right)^2 = 10^{-3}$ dynes per pair of molecules; which is 2000 times as great. Not wholly incomparable therefore; but great enough to give the observed tenacity as a small outstanding or residual phenomenon.

Residual Affinity and Cohesion.

Residual affinity need not be a mere indefinite appellation, it can be precisely defined. It represents the internal force between saturated molecules, holding a group of them together, condensing them into smaller compass than they would otherwise occupy, reducing their tendency to fly away from each other—condensing them in extreme cases into a liquid. It therefore produces the same sort of effect as external pressure produces, and it may be defined as equal to the external pressure which would produce the same condensation or compression on the same number of molecules of a perfect gas. A perfect gas has no residual affinity, its molecules are quite independent of each other. It does not, indeed, accurately obey Boyle's law as usually stated, because the molecules occupy a certain fraction of the whole volume—not all the space occupied is free; but it obeys the law, $p(v-b)=\mathrm{RT}$, where b represents the aggregate molecular bulk, reckoned as cubes not spheres. An imperfect gas or vapour, however, with a residual affinity, K, obeys the law, $(p+\mathrm{K})(v-b)=\mathrm{RT}$. And this K is known as "Laplace's molecular pressure" or internal force. But K is not constant, it increases with the square of the concentration; hence by Van der Waals it was written as $\frac{a}{v^2}$, thus giving us the famous and well-known Van der Waals' expression, which more or less approximately embodies the properties of gases, vapours, and liquids. A measure of K can be made by compression experiments, though of course in any substance for which K is great the compressibility is small. A first approximation is the incompressibility or elasticity of volume, that is $-vdp/dv$; but this is mixed up with the quantity b, and so K is not to be disentangled by incompressibility alone. A better mode of estimating it is by the energy of disgregation of a liquid: in other words, by the work that is needed to pull the molecules asunder from their cohesive attraction—that molecular attraction which has liquefied the substance—and to scatter them as vapour:—in other words by determining the latent heat of evaporation. For instance, at $t°$ C. this is well known to be for water $607 - 0{\cdot}7t$ calories per grm. and interpreting this in energy units by multiplying by J, (that is by $4{\cdot}2 \times 10^7$,) it becomes, for water near the ordinary boiling point, $540 \times 42 \times 10^6 = 23{,}000 \times 10^6$ ergs per c.c., that is $23{,}000 \times 10^6$ dynes per square centimetre or 23,000 atmospheres, or a quantity of that order of magnitude. No wonder then that there should be both chemical and physical evidence that in the liquid state water molecules are liable to cling together into prodigious aggregates, and that the substance has a remarkably high specific inductive capacity.

An increase of molecular cohesion tends to raise the boiling point, and likewise the surface tension, as in the case of mercury. The critical point likewise depends largely upon the value of K; and if K is zero liquefaction is impossible.

The smallness of residual affinity between the molecules of helium and other members of the argon group of gases explains why they so strongly resist liquefaction; but whether they liquefy or solidify first, depends of course merely on whether their boiling point is above the freezing point or not. In a vacuum, the freezing point of water is above its boiling point, and accordingly, in a vacuum, ice only evaporates and does not liquefy: an observation made by Carnelley. To liquefy helium the pressure would have to be increased; though whether it would be possible to do it then remains to be seen. It would be more easy to try the experiment first with neon. (See A. W. Porter, "Nature," 12 March, 1908.)

But how can we measure the residual affinity or molecular cohesion for a thing not in the liquid or the solid form? How can we measure it for a nearly perfect gas? Joule and Thomson devised a way. The gas is cooled and compressed as much as possible, so as to give the molecular attraction every chance; and then suddenly let go, the cold of expansion being observed. If the pressure could really be reduced suddenly to zero there *need* be no cooling, because then no work of expansion would have to be done. All that would then be observed would be the residual cooling due to the molecular attraction. But as this is impracticable, the plan adopted must be to compare the cold of expansion with the heat of compression, and see if they are the same. If, when a gas expands say to double its volume without doing any work, the temperature is the same after expansion as before—when local eddies and rushes have subsided—the gas is a perfect gas, and Laplace's K or Van der Waals' $\frac{A}{v^2}$ is zero. For such gases as hydrogen and helium this is very nearly true, but not quite; and it is this residual affinity or molecular cohesion which in the last resort must be depended on to liquefy or solidify them. All liquefaction depends upon this term, and upon nothing else. External pressure cannot do it, for directly that is relieved the compressed gas completely recovers. But if there is any trace of molecular cohesion, then it exhibits signs of imperfect elasticity, or a more or less permanent set. It is most noticeable when the concentration is excessive, for, as said before, it depends on the square of the concentration, $K = A\rho^2$.

This is the basis of Lord Kelvin's determination of absolute zero. A perfect gas without cohesion would obey the law $p(v-b)=RT$, and the temperature at which its pressure became zero would be the absolute zero. At any temperature above that its molecules would have some movement, and would accordingly by bombardment exert some pressure.

The absolute temperature would be the reciprocal of its coefficient of expansion per degree at constant volume, $T = p\frac{dT}{dp}$, and could be so measured. But as no such gas exists, experiments have perforce to be made on an imperfect gas, and a correction applied for the amount of its imperfection. In this way Joule and Thomson determined that the absolute zero lay between 273° and 274° C. below the freezing point of ordinary water—more exactly it came out 273·2°—a determination which has a real and living interest now that Dewar and others are constantly working down to within a small number of degrees above this ultimate limit, below which it is impossible to go.

It must be for ever impossible, because at that temperature bodies possess no heat, and therefore cannot have any removed. I do not say that the atoms possess no energy—far from it—their energy of internal constitution need not necessarily have been affected; but they possess none of the unorganised energy of molecular motion which we call heat: each molecule, as a whole, at absolute zero, is at rest.

Electrical view of Matter.

Our present attempt to see into the structure of the atom is by resolving it into an assemblage of electric charges, most of them of minute specified size, and all of one sign, called electrons; though the complete aggregate must possess also an equal charge of opposite sign, so as on the whole to be neutral, except in so far as an outstanding unit or two may convert it into a monad or a dyad ion. How the negative charges are arranged, and what form the equal positive charge or charges may take, are problems still awaiting solution, to complete this part of the theory of matter.

Comparatively early speculation and calculation on this subject are contained in a remarkable paper by Lord Kelvin, published in the Phil. Mag., 1901, under the title "Æpinus atomised." It is reproduced in the volume of Baltimore Lectures as Appendix E. To him also we owe the first attempt to work out the statics of an atom according to a simple conception whose major consequences can be traced with comparative ease: viz., that of a spherical portion of uniform positive electricity in which minute negative charges are sown like specks: being attracted towards the centre of the sphere according to the law of direct distance, and repelling each other according to the inverse square law. (See Fig. 1 below.)

Various interesting groupings can be easily calculated as stable patterns; and they tend to become more stable if they revolve regularly round the centre or if sustained by a central group or particle. But although the distribution and properties of such groups is profoundly suggestive and interesting, the behaviour of atoms consisting of a mass of positive electricity with even a single electron inside them became instructive in the hands of Lord Kelvin.

Thus equal and similar neutral atoms of this kind (the positive charge equalling the negative in quantity, not in size), would exert no force on each other, *till* they touched and partially interpenetrated: then they would attract and cohere. But they would not presumably interfere with or damage each other, and when separated would be as neutral as before. But if one sphere were distinctly smaller than the other, though equally charged, the neutrality would not continue. The more concentrated would be the more powerful. Pressed together the electron at the centre of the larger one would be much displaced, and under certain conditions would actually be absorbed and retained by the smaller sphere, so as to be carried away by it when they separated again. Thus, after the collision, one of the atoms would have become positive, by losing its negative electron; and the other would be negative, by having acquired one too many. The likeness exhibited, by the phenomena of such a presentation or model, to contact electricity, and to the strong electrification produced between insulators when pressed or rubbed together is obvious.

A sphere of positive without negative nucleus, brought near or in contact with an equal sphere possessing one. would not be able to rob it; nor could a big one ever take the nucleus from a smaller one. But one which was even a little smaller, any size less than 0·89 the diameter of the bigger one, would, if it approached close enough, pull the electron completely out and assimilate it itself.

It is confessedly unlikely that any atom should have only one electron; but that which it is fairly easy (not too easy) to calculate for this simple case would happen in modified fashion in more complex cases, when the electrons were more numerous; and a vast number of possibilities suggest themselves, such as with ingenuity can be made to account for a great variety of the fundamental facts of chemistry. Ingenuity such as is needed for this difficult work, which at present is only in its infancy, has been conspicuously displayed in various well-known writings of Professor J. J. Thomson.

Meanwhile to return to Lord Kelvin:—Where there

are more than one or two electrons in the atom, constituted as supposed above, they distribute themselves in stable form as follows:—

Three form a plane equilateral triangle.

Four are at the corners of a regular tetrahedron.

Five may be stable if four are in a plane at the corners of a square and one centrally perpendicular to the plane of the square; but they are more certainly stable if three are in a plane, and two are on its central perpendicular, one on each side.

Six are stable at the corners of a regular octohedron; they are also possibly stable in a plane if one is at the centre and five at the corners of a pentagon.

Eight can be at the corners of a cube, but with insecure stability, unless there are some at the centre too.

Twenty, at the corners of a regular dodecahedron, would either have to be *constrained* to lie on a spherical surface or else would have to be well sustained by repulsive forces from the interior.

The calculation for space-distribution becomes complicated, and most attention has been paid to distribution in a plane: wherein the criteria for stability, as depending either on central repulsion or on centrifugal force due to regular rotation, has been admirably worked out by J. J. Thomson in a methodical and very instructive way. A typically instructive and simple case is that of four electrons revolving round the centre. So long as they are revolving above a certain angular speed they are stable when at the corners of a square; but if the speed falls below a critical value, this grouping becomes unstable and they fall, with a convulsion and emission of energy, into a statical position of equilibrium at the corners of a tetrahedron. It is in a direction typefied by this simple example that some of us are looking for an explanation of spontaneous radioactivity.

Theory of Electrons.

The whole idea of being possibly able to express matter in terms of electricity is, however, one that needs reiterated explanation and emphasis. The mode in which the electrons themselves have been weighed and counted, I will not here again enter into. It may be taken as known; and if not, it can be found described in many places—among others in my book called "Electrons."

But the fundamental fact on which is based the notion that matter may be an electric phenomenon, is deserving of more expanded and repeated notice; and part of this lecture must be devoted to that. For to the superficial eye nothing could be more distinct than matter on the one hand and electricity on the other. Anyone would suppose that an electric charge, whatever it is, must be a mere appanage or decoration of matter, and that by no conceivable ingenuity could it be thought of as constituting the substance of matter itself. Benjamin Franklin—whose ideas concerning electricity have in many respects become so singularly justified—would, I conjecture, have been profoundly interested, but somewhat astonished and probably sceptical, if such an idea could in his day have been authoritatively mooted. If so, it would be one of the many instances when ultra and exaggerated scepticism concerning truth and reality fails to be justified by events. Unless indeed even the present tendency be discounted, and sceptics prefer to wait till future times for a more thorough justification of these theories before even tentatively accepting them: —a legitimate though not a prolific or fertile attitude.

Electric Mass.

The attribution of the material property of "mass" or *inertia* to electricity entirely depends upon the magnetic properties of an electric charge in motion. *The magnetic properties of an electric charge in motion.* In that sentence I have mentioned the three fundamental attributes, or manifestations, or phenomena, of the material universe:—Electric charge, Magnetic field, and Mechanical motion; or Electric force, Magnetic force, and Mechanical force. These three vectors are essentially perpendicular to each other, and correspond to the three dimensions of space. They also correspond to the three chief known properties of the ether of space, which is the substratum or material basis of all activity. We do not yet know the constitution of this continuous and omnipresent medium, but it is certainly so constituted that each of these three vectors is physically possible at any point in space; and moreover it is so constituted that wherever two of them exist, the third springs into being—at right-angles to both.

A charge in motion constitutes a current—"current" is our shorthand expression for charge in locomotion—

and a current is always and essentially surrounded by a magnetic field. A magnet in motion generates a current in conductors, and an electromotive force in insulators. A varying magnetic field does the same, for its lines are then in lateral motion. A magnetic pole in the neighbourhood of an electric charge, even when stationary, sets up a theoretical spin in the ether, round the line joining them—a hypothetical spin which can display itself by its gyrostatic action.

The result is that when a charge is in motion, the magnetic field due to the motion, when once generated, tends to preserve and prolong that motion, so that mechanical force is needed to obstruct or stop the moving charge. Similarly to accelerate or set in motion any electric charge involves the generation of a magnetic field, and therefore requires the application of mechanical force and the expenditure of energy. No such force is necessary so long as the motion is steady, no more than if the motion was zero; but directly the motion changes, whether in the way of increase or of decrease, force is immediately exerted: nor can there be any kind of change produced in the motion, either in magnitude or direction, save by the exertion of force. In other words, an electric charge in motion, not acted on by any force, will continue to move in a straight line with a uniform velocity. And if a force acts upon it, the acceleration produced is exactly proportional to the force.

Thus the reaction between electric and magnetic fields precisely simulates all the effects of mechanical or Newtonian inertia. It is an acceleration-coefficient, not a velocity-coefficient; it is not of the nature of resistance or viscosity but it is of the nature of mass. But when I say simulates—I would rather say that it accounts for, or rationalises, or carries one step further on, the otherwise blank and bald idea of Newtonian inertia.

The first two of Newton's Laws of Motion are thus obeyed by electricity. Is the third law obeyed likewise? I say yes; but the answer is not so simple and easy. The peculiar condition of an advancing wave-front has to be taken into account, and we have to admit that an ethereal wave advancing with the speed of light may possess attributes which hitherto have been thought to belong only to matter, and may—indeed must—sustain the reaction of a stress whose other end acts upon the source. So that light or radiation of any kind must exert a pressure and must be the seat of a stress;—a fact which, predicted by Maxwell, has now indeed been directly and experimentally observed.

If somehow there should be acceleration without the action of any material body, then waves are generated in the ether; and the mechanical force essential to the action is supplied by the reaction of the train of ether waves. It is a singular but very important case: the wave-front seems to represent adequately some of the properties of matter, and the generation of such a wave-front is the inevitable consequence of either accelerated or retarded or curved motion of an electric charge. This and this alone is the source of ethereal radiation.

To pursue this subject would lead us into the intricacies of spectrum analysis, which has already begun to throw light upon atomic constitution, but what little I propose to say on that subject will come better later.

Leaving this as a digression, and returning to electric inertia, it may be truly said that an electric charge has the property which we know as mass or inertia; and it becomes a possibility that no other mass or inertia need or does exist. This is the chief basis of the electric theory of the nature of matter. It is one of the foundations on which the notion that an atom of matter is built of electric charges must fundamentally rest.

But, admittedly, the extra mass of any ordinary charged body, due to the charge, is excessively and absurdly small. The mass depends not only on the charge but also on its concentration; in other words, on the charge and its potential. It is in fact proportional to the electrostatic energy of the charge. The mass or inertia of an electric charge is such that, if moving with the velocity of light, its kinetic energy would equal the electrostatic energy of the charge at rest. In other words, considering the charge uniformly distributed through a sphere of radius, a, $\frac{1}{2}mv^2 = \frac{1}{3}e\frac{e}{ka}$; but it is known that $v^2 = \frac{1}{\mu k}$, so $m = \frac{2}{3}\frac{\mu e^2}{a}$, a simple formula which is the sheet-anchor of the electrical theory of matter.

To build up an atom out of a reasonable number of electrons, therefore, the charge of each electron must be excessively concentrated, that is to say, must occupy exceedingly small bulk, so that an electron's intrinsic potential may be high—comparable indeed to a million volts: an astonishing value, which reduces itself to only a few volts at any distance comparable with molecular dimensions. It is indeed possible to specify the bulk of an electron which will enable us to say that its mass is wholly of the electrical kind. Every experiment made—both direct and indirect—goes to confirm the extraordinary smallness and consequent penetrating power of electrons.

At one time, not so long ago, we were in doubt about the size of atoms; some few may perhaps have been in doubt as to whether such a phrase had a legitimate meaning. I hope that this unnecessary scepticism is now set at rest, and that everybody admits the estimate of size or molecular magnitude which we owe to the genius of Lord Kelvin, among others. The extreme minuteness of the atom can be brought home to the imagination in many ways; but one striking way is to point out that in spite of the extreme rarity or dilution of the gold chloride dissolved in sea water—estimated at 1/50th grain per ton, or say one part in a thousand-million—so that it appears not to be possible to extract it economically—yet that in every cubic millimetre drop of that liquid, the number of gold atoms amounts to several hundred millions. And the number of atoms of radium in any 50 c.c., or say a half pint, of sea water* or of ordinary rock is about the same.

Yet in spite of their excessive minuteness we know of one definite structure which can only be about ten or a dozen molecules thick—namely the soap bubble on the point of bursting, in the region which shows itself to the eye as a black or non-reflective spot.

Waves of light are excessively small things, of which a hundred thousand only span an inch, and yet waves of light are coarse unwieldy things compared with atoms; so that the fineness of microscopic vision must break down in the attempt actually to see the atoms, although by more or less indirect means there is some hope that the larger kind of molecular aggregates may some day become in a manner visible, by the diffraction and other effects which they cause in a beam of light. Moreover, the excessively thin ether pulse known as an X-ray appears to be distinctly thinner than molecular dimensions, since it is able to pass through a mass of material with only a moderate amount of loss.

*Joly estimates the proportion of radium in sea water as 2×10^{-14} *Phil. Mag.*, March, 1908.

But although the atom is so minute, it is huge compared with an electron; and if it be true, in any sense, that an atom is mainly composed of electrons, they must be as far apart from each other inside the atom, in proportion to their size, as the planets of the solar system are in proportion to their size.

The porosity of matter so constituted is extreme, and the ultimate specks of which even an atom is composed are as it were few and far between. Though an atom may perhaps contain thousands altogether, it consists mainly of intervening space in which there is nothing but continuous unmodified ether.

That is the kind of picture which we are beginning to form in our minds, not dogmatically but as a guide to investigation; it forms a mental image of which we can trace some of the consequences, so as to be able to check and amend it by observation and experiment.

It is on this basis that I have been led to attribute so excessive a density or massiveness to the ether—that is to say, on the supposition that it is of ether that the electrons are really composed, that they are, as it were, "singularities" or some kind of structure or strain or motion-centre in the ether, differing from it only in the circumstance of their arrangement, not in their density or concentration. A really continuous fundamental medium cannot be either rarefied or condensed: it must be uniformly everywhere. Accordingly, whatever density a discrete aggregate of scattered particles may possess, the density of the continuous medium, which composes them and fills up all the intervening spaces, must be greater, and may be immensely greater.

The reason why matter appeals to us so clearly and strongly is because our present bodies are composed of it, and because our sense-organs have been evolved to respond to its various motions; but matter turns out to be a mere faint misty modification of the great medium itself; and, compared with it, is like a gauzy cobweb or mist or a milky way. I repeat this because I believe it will be found to be a fact of profound significance—a significance as yet undreamed of—significance which could only at present be dealt with in terms of speculation, a thing desirable to avoid until we can adduce a more substantial foundation for it.

Electron History.

It may be convenient now to trace some stages in the history of the discovery of the electron, or separate physical unit of electricity: that is to say stages in the discovery that electricity is not a continuous medium, but consists of particles or atoms—truly indivisible particles so far as we know at present—and that an ordinary electric charge, as we know it experimentally, is due to either the gain or the loss by a body of some of those particles; also that the particles can exist in some cases apart from matter, either in the interstices between the atoms, or in free space away from matter altogether.

The first inkling of the discovery of the atomic nature of electricity dates back to Faraday and his experiments in electrolysis. The indivisible atomic charge—the charge of a monad ion—was then practically measured, at least as soon as the size of the atom was known; and an "atom of electricity" was spoken of, both by Maxwell and by Helmholtz. In amount, measurement indicates that it consists of $3{\cdot}4 \times 10^{-10}$ electrostatic unit, some uncertainty affecting the second significant figure. On this basis, Dr. Johnstone Storey gave to these unit charges their name "electron," and speculated ingeniously on some of their properties. Then J. J. Thomson found that in the rarefied air of a vacuum tube these charges were loose, flying as cathode rays; and identically the same, no matter what was the kind of residual matter in the tube. It was in this state that their mass was measured, and found to be a thousand times less than the previous material minimum—the hydrogen atom. Then Zeeman found that their orbital motion was the source of all bright line spectra—that is to say, that they emitted radiation, carving the ether into waves.

The conduction of electricity is due to the procession of electrons and positive charges; and the good conducting power of metals, whether for heat or for electricity, is supposed to be due to a crowd of electrons freely permeating the interstices between the atoms, being handed on from one to another so readily that they are practically dissociated or free in vast numbers. And the long-known connexion between heat conductivity and electric conductivity is found to be thus rationally and quantitatively accounted for.

Furthermore, the regular gradation of electric properties exhibited by the elements in Mendeléef's classification can be stated electrically, in a way that certainly suggests an approximation to the truth. Each octave has an electro-positive and an electro-negative end, the most extreme members being at the ends, and an almost neutral body in the middle, of each series ; and in the step from the extreme electro-positive of one octave to the extreme electro-negative in the next, an intermediate halting place of a neutral and quite inert body is now known, by the discovery of the argon-krypton-xenon series. (See diagram 4 below.)

The atoms of the electro-negative set are those which can easily make room for, and have acquired, extra electrons—more than properly belong to their constitution ; these are the non-metals, and are chemically active on that account. The atoms of the electro-positive variety are those which easily lose, and have lost, some of their electrons—thereby becoming good conductors by reason of the loose electrons which they have let go ; these are the metals. A strong electric attraction naturally exists between members of the electro-negative and the electro-positive group, because they are oppositely charged ; and the clinging together of these atoms, under electric attraction, constitutes chemical combination. Conducting power usually disappears after combination, except the variety of it which may be brought about by the dissociation or breaking up of the compound molecules.

The epithets positive and negative, which from the electrical point of view seem so interchanged and inconvenient, from the material point of view get right again. For the atom which is hungry for more electrons and has absorbed them is electro-negative ; while the atom which has a surplus and has extruded some is electro-positive. A tetrad can be either the one or the other. In methane the carbon is presumably acting as a negatively charged body. In carbon tetrachloride the carbon is presumably acting as a positive. Chlorine seems to have much more residual affinity than hydrogen has, and accordingly the molecules tend to cling together, and the substance is a liquid instead of a gas.

Once more I repeat that the term in Van der Waal's famous formula, $\frac{a}{v^2}$, or Laplace's K, is the liquefying force, and is a measure of this outstanding affinity ; it is determined by the latent heat of evaporation, or to some extent by the boiling point and the critical point.

Spectra.

The critical and numerical examination of a bright-line spectrum, both when free and when under magnetic influence, is one of the most powerful means of detecting the internal structure of an atom; and although the subject is a difficult one, considerable progress in this direction is evidently going to be made: indeed, has already been more than begun. There are, however, a good many outstanding difficulties and some puzzles to be overcome, showing that our ideas, though probably true as far as they go, are as yet incomplete and require either modifying or supplementing; or most likely both.

Another means of examination, likewise dependent on their ethereal connection, is the mode in which they absorb and disperse light. Definite or bright-line spectra are now supposed to be due to the constituent electrons of an atom; while confused or continuous spectra are supposed to be due to more general dissociated electrons, with no individual atomic constitution or precise geometrical grouping. Bright-line spectra are to continuous spectra somewhat as polarised light is to common light—a more definite accurate precise metrical thing. One result of the theory, and one that could be tested, is that any substance that gives a bright-line spectrum ought to be a non-conductor. And *vice versâ*, as long as a thing is a non-conductor it should give a bright-line spectrum. Any trace of a continuous spectrum must be associated with some amount of electric conductivity, that is to say, with the production of loose or dissociated electrons in addition to the tight or continuous ones.

Gases and many vapours, as is well known, are in the strict sense non-conductors, and their spectrum does consist of bright lines.

The effect of a magnetic field applied to a bright-line source, in tripling or otherwise complicating the lines and causing the constituents to emit definitely polarised light, has been triumphantly accounted for, and was indeed, in part predicted, by the electron theory as elaborated by Lorentz. The explanation is so complete and satisfactory that there can be no reasonable doubt of its truth; and experiments made on those lines are leading to a kind of atomic astronomy, whereby the orbits of electrons and their perturbations can be examined into and mathematically expressed,—a theory begun some time ago by Johnstone Storey.

The gradation of the more specially chemical properties, exhibited by each octave of Mendeléef's series, can also be incipiently accounted for by the electric theory: the valency increasing from 1 up to 4 in the middle of the octave, and then decreasing again down to —1. And the peculiar properties of the quadrivalent atom, whereby it can cling together in groups, and give all the compounds of organic chemistry, comes out as a rather natural property. (See diagram 4 below.)

Even the peculiarities observed in the spectrum of the different elements in each group show some indication of yielding to a rational theory; for, as has been already said, mechanical stability requires that the elements shall arrange themselves in rings, with a given number of electrons in each ring. And wherever a ring of three or of any other fixed number occurs, it is natural to suppose that certain properties, especially certain radiational properties, will be associated with that particular number. This same number will occur in the corresponding member of higher and higher groups—fresh rings being added as the atomic weight increases—but the numbers in the inner rings will repeat themselves in octaves, so as to reproduce something of the old properties, together with new properties in addition, due to the extra complexity. (See diagrams 2, 3, and 4.)

It should be understood that the visible and plentiful radiation which furnishes a spectrum can hardly be due to the gentle emission of waves from isolated and free atoms in regular revolution: some violence is needed to generate radiation in conspicuous amount. They require to be shaken or jogged; just as a bell does not sound when merely waved about—it requires to be struck. *Sufficient* acceleration will always make it sound. Plentiful radiation is only emitted, therefore, when atoms clash together, as they do when they chemically combine, or as they may be made to do by that considerable quickening of their motion which we call a great rise of temperature.

Visible light seems to be due to the banging and clanging of molecules, much as X-rays are due to the violent impact of electrons on a target. The system is then shaken, and the atom emits for a short time its characteristic waves. The stopped electron only emits a single thin shell.

Atomic Structure.

But now returning to the actual internal structure of the atom—is there anything definite that we can say concerning that? Assuming that the mass of the atom is due to an assemblage of electrons, can we say anything clear and distinct concerning the way in which those electrons are arranged, and can we descry what they are doing among themselves? First of all, it is certain that the arrangement is not stagnant or quiescent: it must be full of internal energy. Lord Kelvin, indeed, surmised that the energy was statical, akin to the energy of gunpowder or a compressed spring. To me and others it seems far more likely that it is kinetic, analogous to the energy of a flywheel or of the planets in their orbits. But that there is a great store of intra-atomic energy we know, from the fact that some atoms every now and then explode and fire off a projectile from time to time, just as a loaded and repeating gun may fire off a cannon ball or a bullet. This energy may, as hinted, be due to motion already existing among the parts of the atom, or it may be due to some unexplained static force held locked until some trigger or detent is pulled, and then suddenly liberated. Both are possible modes of statement, but the latter evidently leaves more in a state requiring further explanation than the former does. Some, as we have said, think that the energy is static or potential, others that it is more likely to be kinetic: but whatever the nature of the energy, there is no valid doubt as to its amount. And it is enormous. Not so great as, nor even comparable with, the energy in the same bulk of the ether of space, but still enormous, and entirely eclipsing the insignificant amount corresponding to chemical combination, which is all that we have so far learnt to utilise. The energy of heavy atoms is greater than the energy of light atoms, in proportion to their atomic weight, but the intra-atomic energy stored up inside the atoms of a gramme even of hydrogen is 10^{19} ergs. That is to say, is enough to raise a weight of a thousand tons 60 miles high; or, otherwise expressed, enough to convert 300 tons of ice into high-pressure steam. Even a slight rearrangement of constitution, emitting only a minute fraction of this energy in the process, may therefore be depended on to give powerful radiation, and amply accounts for the phenomena exhibited by radium and its congeners.

The occasional disruption or explosive break-up of an atom, can readily be stated in terms of the electric theory; and the fact that a revolving electron must radiate away its energy, thereby reducing the motion of one of the rings sooner or later to a condition of instability, must necessarily lead at intervals to a catastrophe and rearrangement—the rearrangement being perceptible to us as the appearance of a different element, and the catastrophe making itself apparent by ejection of some discarded constituent, with all the phenomena of radioactivity.

Not that the theory of radioactivity is at all as yet complete; but the lines on which an explanation is to be sought, and the theory elaborated, are fairly clear.

When I consented to give this lecture in memory of Dr. Hurter, it was with the idea of trying to show how promisingly an incipient explanation of the properties associated with the chemical series of elements had been begun by J. J. Thomson, on the lines of the kinetic hypothesis. But it is hardly possible to give an intelligible account of this matter, by the method of oral exposition, in any reasonable time. Nor is the theory as yet so complete and satisfactory that we h ve any confidence in maintaining it to be certainly a step in the direction of the truth. It may be that it leads off the track—I do not think so—but I could not bring it forward with the same confidence as I bring forward more general and less detailed considerations. Suffice it to say that Thomson has made a heroic attempt to exhibit Mendeléef's series, and the gradation of properties among the so-called elementary bodies known to chemistry, and some of the peculiarities of their spectra, on the lines of a set of revolving rings of electrons obeying mechanical laws and spinning in regular orbits inside a uniform positive charge. (See diagram 3 below.)

Evolution.—The notion of atomic evolution would suggest that complex atoms were formed by the falling together of simpler ones; or by the gradual growth of complexes by assimilation of electrons. This process has not yet been observed. What has been observed is the converse process—the disintegration and gradual collapse of some more massive atoms, and the emission by them of lighter materials. But some indirect evidence that the inverse or constructive process may also occur is furnished by electroscopic examination of the nebulæ. For, just as we regard these as the probable parents of solar systems, so we must notice that the elements of which they are composed are the most simple ones—those of very small atomic weight: Hydrogen, helium, and the like. Hence it is natural to think that just as the cloud contracts into suns and planets, so its atomic materials coalesce into heavier elements.

If it be asked why is nothing known lighter than hydrogen, the answer must perhaps be sought in the comparative instability and short life of any *very* small collection of electrons; for the radiating power of a very few would be considerable, even without collisions. But the most likely answer is that the grouping has acted as a unit and fallen together into aggregates, of which the more stable and permanent ones are recognised by us as the chemical elements. Some of them are not very permanent, however. Polonium, actinium, radium, and others, have a lifetime that can be measured; and even that of uranium can be estimated. Probably in time we shall know something about the genealogy and vitality of the whole chemical population.

Uncertainty and vagueness still hover round the conception of the positive charge. The negative electron or corpuscle is now a fairly familiar object, of the roadside, so to speak; but the positive unit, which is probably best not called an electron, still lurks in holes and corners, and either by reason of sluggishness, or for some other reason, resists being dragged out into the light of day.

Diagram I.

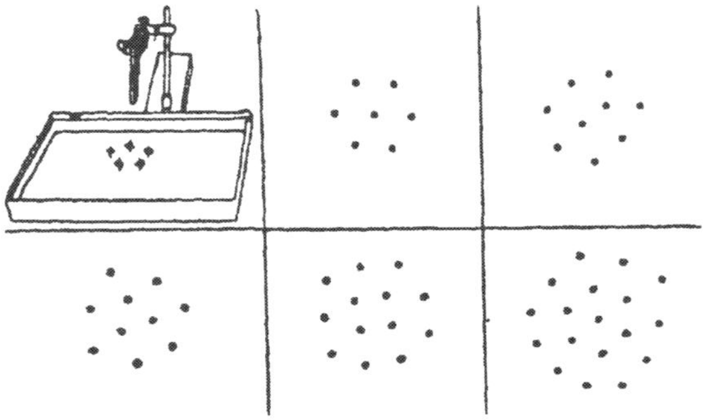

Progress is being made, however, and during the next decade one of the most promising and repaying lines of advance will be a quest into the nature and properties of the unit positive charge.

Diagram 1 shows the arrangement of a number of magnets floating on water and attracted by a large magnet, centrally situated, high up. These roughly imitate the electrons in a sphere of uniform positive electricity, save that they are forcibly constrained to keep on a plane. This artificially increases their stability; but even when they have freedom to move in a third dimension the same amount of stability can be attained, provided they are rapidly revolving—as electrons almost certainly are.

DIAGRAM 2.

Spectrum to be expected from three revolving electrons, in atoms of higher and higher atomic weight.

They have nine degrees of freedom and therefore may be expected to give nine bright lines. But in simple cases some of these could be coincident.

(*a*) Three electrons alone might give ......

(*b*) Perturbed by one outer ring these would become

(*c*) Perturbed by two outer rings they would separate out thus

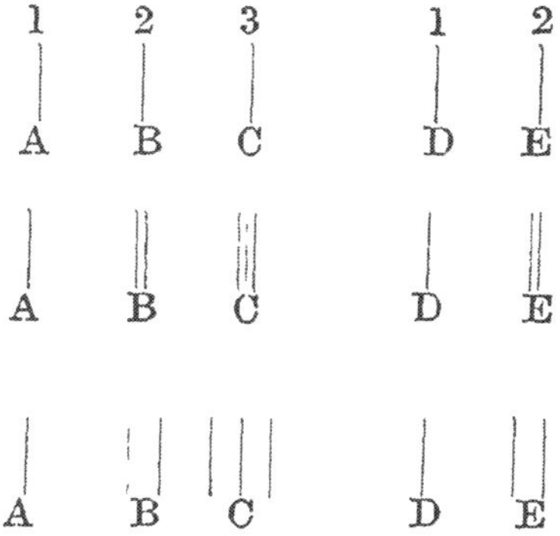

Fresh lines would also be added to correspond with the extra rings.

Diagram 2 shows the spectrum to be expected from a group of three electrons, grouped as an equilateral triangle, which in the aggregate have nine degrees of freedom. But in the simplest case some of the lines will be coincident (corresponding to equal roots of an equation) and accordingly we may suppose that only five lines appear, as in 2*a*, some of which are double and others triple.

Surrounding the triangular group with a large ring of electrons is equivalent to stepping up, by greatly increasing the atomic weight, so as to reach a corresponding member in the next higher series; and the perturbation of the triangular grouping when now set vibrating will resolve the component lines a little, and give Fig. 2*b*.

Stepping to the corresponding member of the next higher series, it gives the same lines because it again contains a triangular group; but now the separation of the lines is more thorough, giving Fig. 2*c*. Other lines are, no doubt, added by the extra rings, but these have no relation to the ones characteristic of the triple group.

DIAGRAM 3.

Hypothetical reason for gradation and similarity of properties.

Approximate substance.	Total number of electrons in atom.	Arrangement in successive rings.
Be	11	3 8
Mg	24	3 8 13
Ca	40	3 8 13 16
Zn	60	3 8 13 16 20
Sr	84	3 8 13 16 20 24
Cd	110	3 8 13 16 20 24 26
Ba	140	3 8 13 16 20 24 26 30

Diagram 3 shows a set of hypothetical rings, such as may readily occur, corresponding to a complete Mendelejeff family of higher and higher atomic weight. The similarity of properties is emphasised by the repetition of the inner rings; the rise of atomic weight and corresponding modification of properties is a consequence of the extra and outer ring. The whole table, and the substances specified in the first column, are only here to be regarded as typical or suggestive of the kind of thing that may be looked for, not as a statement of what has been found.

DIAGRAM 4.

Typical group, or Mendelejeff family, on electric theory.

Type of Substance.		Total number of Electrons in atom.	Arrangement in Successive rings.	Valency.
? H		58	19 16 13 8 2	– i
He	Ne	59	20 16 13 8 2	0
Li	Na	60	20 16 13 8 3	+ i or – vii
Be	Mg	60	20 16 13 9 3	+ ii or – vi
B	Al	62	20 17 14 9 3	+ iii or – v
C	Si	63	20 17 13 10 3	+ iv or – iv
N	P	64	20 17 13 10 4	– iii or + v
O	S	65	20 17 14 10 4	– ii or + vi
F	Cl	66	20 17 14 10 5	– i or + vii
Ne	A	67	20 17 15 10 5	0
Na	K	68	21 17 15 10 5	+ i
Mg	Ca	69	21 18 15 10 5	+ ii
&c.				&c.

Diagram 4 exhibits the converse arrangement: showing in a typical or representative manner all the neighbouring elements of a single Mendelejeff series—the one welding link being the possession of twenty in the outer ring. Among those at the top of the list, the 20 is barely stable and is liable to shed one of its electrons, becoming thereby positively charged. Those at the bottom could easily take another on without difficulty, and often do—becoming negatively charged.

The valency exhibits a regular gradation from +1 through +2, +3, ±4, to –3, –2, –1. And the extremes of the series have the valency 0. A permanent change in the number of the outer ring carries one over into the next series and the valency suddenly changes from the active valency, –1, to the equally active valency, +1; stepping, as we now know, through 0 as an intermediate step over the discontinuity.

This hasty treatment by no means does justice to the whole conception, but it may suffice as a hint or indication, as much as at present is legitimate; and it may have the effect of sending students to the original sources which he must read with critical appreciation but without slavish acceptance.

The substance indicated is only typical, and either of the first two columns may be taken as representing the group typified. The second column really comes below the first, and leads on a third, etc. They are put together in the table for compactness. The total number of electrons given does not correspond to any particular element necessarily, but this number would adjust itself in the rings shown. The actual total number may be much greater.

Appendix on absolute temperature.

The elementary argument about the definition of absolute temperature in terms of a perfect gas can be put thus :—A perfect gas is one whose molecules act on each other, and on the walls of the containing vessel, solely by bombardment. Simple mechanics shows that such a substance exerts a pressure

$$p = \tfrac{1}{3}\rho u^2 \;; \quad \ldots\ldots\ldots\ldots \quad (1)$$

and whenever it expands, all the work done is against external pressure. The heat in such a body is solely the energy of its irregular or unorganised molecular motion; and the temperature of such a body can be defined as simply proportional to the heat, or equal to the heat divided by a capacity-constant, mc.

If the gas has to expand against external pressure more heat must be supplied to allow for the external work done, $\int p\, d\, v$; the capacity being now called mc' if the pressure is constant. Consequently, if the gas be heated at constant pressure from absolute zero up to the temperature T, the heat required can be expressed as $\mathrm{H} = mc\mathrm{T} + pv = mc'\mathrm{T}$; wherefore—

$$p = \rho(c' - c)\mathrm{T} \quad \ldots\ldots\ldots\ldots \quad (2)$$

which may be called the characteristic equation of the substance. Comparing this with the first equation, we see that—

$$u^2 = 3(c' - c)\mathrm{T} \quad \ldots\ldots\ldots\ldots \quad (3)$$

which constitutes a definition of absolute temperature in terms of the characteristic constant, $c' - c$; the "3" having reference to the three dimensions of space.

Actually to determine T we can employ equation (2) and get rid of the constant say by measuring the increase of pressure when the gas is heated at constant volume.

This gives $\frac{dp}{p} = \frac{dT}{T}$, or T = $\frac{dT}{dp} = \frac{1}{a}$ .... (4)

the reciprocal of the coefficient of expansion. In other words the expansibility of a perfect gas is simply the reciprocal of its absolute temperature.

This is consistent with the form of characteristic equation which allows for molecular bulk, though not for molecular forces, $p(v-b)=RT$.

For a slightly imperfect gas there is the cohesion or molecular-attraction term to be attended to as well, and its characteristic equation is $(p+K)(v-b)=RT$, which for constant volume warming gives

$$\frac{dp}{p+K} = \frac{dT}{T}$$

$$\text{or } T = p\frac{dT}{dp} + K\frac{dT}{dp}$$

or $T = \frac{1}{a}\left(1 + \frac{K}{p}\right)$ .... (5)

where a is the coefficient of expansion as measured on a constant-volume thermometer, showing that a small correction-factor must be applied, depending on the incipient cohesion or intermolecular attraction, represented by Laplace's K or Van der Waals' $A\rho^2$.

To get K we must perform an irreversible operation, say a sudden expansion, δv, under adiabatic conditions, and observe the resulting small change of temperature δT. If this operation is performed in a rigid vessel, so that no external work is done, the cooling will be due to internal work only; and the energy change can be expressed either as $K\delta v$ or as $mc'\delta T$, when eddies have subsided.

But instead of δv we may write $-\frac{v}{p}\delta p$ since the temperature is nearly constant, so that $K = -\rho c' p.\frac{\delta T}{\delta p}$ (6)

Hence, denoting by θ the small observed change of temperature corresponding to the change of pressure, π, and substituting in equation (5), we get finally as an expression for the absolute temperature of the gas experimented on, $T = \frac{1}{a}\left(1 - c'\rho\frac{\theta}{\pi}\right)$ ....(7)

To interpret this numerically:—The quantity $c'\rho$ will be recognised as the atomic heat, which is nearly the same constant for all ordinary gases, and equal in c.g.s. energy units to
$0{\cdot}2375 \times 0{\cdot}001293 \times (42 \times 10^6) = 0{\cdot}001294 \times 10^6$ ergs per c.c.

The observed change of temperature per atmosphere, observed as the final result of the irreversible Joule and Thomson expansion, was, for air, about a fifth of a degree, or more exactly 0·208°; so that

$$\frac{\theta}{\pi} = \frac{0{\cdot}208}{10^6 \text{ degrees per sq. cm.}}$$

Hence the dimensions of the whole term, $c'\rho\ \theta/\pi$, come out right, as a pure number (being in fact obviously a ratio of two energies when ρ is written m/v); and for air the correction factor equals $1 + 0{\cdot}001294 \times 0{\cdot}208 = 1{\cdot}00027$.

At zero centigrade the expansibility of air was measured by Regnault as 0·0036706, wherefore the absolute temperature corresponding to zero centigrade is, in accordance with equation (7),

$$\frac{1{\cdot}00027}{0{\cdot}0036706} = 273{\cdot}17.$$

Made in the USA
Monee, IL
04 November 2026

67388677R10020